FORCES

By
Robin Twiddy

BookLife
PUBLISHING

Written by:
Robin Twiddy

Edited by:
Madeline Tyler

Designed by:
Amy Li

ISBN: 978-1-80155-824-2
ISBN: 978-1-80155-855-6

*All facts, statistics, web
addresses and URLs in this
book were verified as valid
and accurate at time of
writing.*
*No responsibility for any
changes to external websites
or references can be
accepted by either the author
or publisher.*

IMAGE CREDITS

CONTENTS

Words that look like **this** can be found in the glossary on page 24.

WHAT ARE FORCES?

Forces make things move. There are many different forces. We use lots of them every day.

Two types of force we use all the time are pushing and pulling forces.
This child is using a pulling force to move this dog in a wagon.

PUSH

When we push, we use our bodies to move something away from us.

6

7

JUMP

When you jump, you use your legs to push your body away from the ground.

If you use more force to push, you will jump higher.

PULL
When we pull, we are using a pulling force to move something closer to us.
These kids are pulling together.

When we pull, we can change the speed and direction of something.
Pulling harder or softer will change the speed of the thing you are pulling.
Pull
11

CLIMBING

Have you ever tried to climb a rope? You need to pull hard to make your body go up.

This boy is pulling with his arms and pushing with his legs to climb the climbing wall.

13

SPEED

You can use force to make something speed up or slow down.

Force can be used to turn a bike by pushing and pulling on the handlebars.

15

SCOOTERS

Do you know what forces you are using when you ride on a scooter?

You push on the ground to move the scooter forward.

INVISIBLE FORCES

We cannot always see the things that are making the forces around us.

Wind can push with a very strong force.
Wind turbines are moved around by
the pushing force of the wind.

GRAVITY

Gravity is a force that you cannot see, but you can see how it works all the time.

Gravity is a force that pulls down.

When you jump up on a trampoline, gravity pulls you back down.

FORCES ALL AROUND

There are forces all around us all the time and we use forces in everything we do.

There are many more forces to learn about.
What kinds of forces have you used today?
23

GLOSSARY

DIRECTION	the way that someone or something is moving, for example, right, left, up or down
EARTH	the planet that we live on
SPEED	how fast or slow something is moving

INDEX